Made in United States
North Haven, CT
01 November 2021

10777040R00057